SpringerBriefs in Public Health

SpringerBriefs in Public Health present concise summaries of cutting-edge research and practical applications from across the entire field of public health, with contributions from medicine, bioethics, health economics, public policy, biostatistics, and sociology.

The focus of the series is to highlight current topics in public health of interest to a global audience, including health care policy; social determinants of health; health issues in developing countries; new research methods; chronic and infectious disease epidemics; and innovative health interventions.

Featuring compact volumes of 50 to 125 pages, the series covers a range of content from professional to academic. Possible volumes in the series may consist of timely reports of state-of-the art analytical techniques, reports from the field, snapshots of hot and/or emerging topics, elaborated theses, literature reviews, and in-depth case studies. Both solicited and unsolicited manuscripts are considered for publication in this series.

Briefs are published as part of Springer's eBook collection, with millions of users worldwide. In addition, Briefs are available for individual print and electronic purchase.

Briefs are characterized by fast, global electronic dissemination, standard publishing contracts, easy-to-use manuscript preparation and formatting guidelines, and expedited production schedules. We aim for publication 8-12 weeks after acceptance.

More information about this series at http://www.springer.com/series/10138

Khalid Rehman Hakeem
Waseem Mohammed Abdul
Mohd Muzzammil Hussain
Syed Shoeb Iqbal Razvi

Oral Health and Herbal Medicine

 Springer

Khalid Rehman Hakeem
Department of Biological Sciences
King Abdulaziz University
Jeddah, Saudi Arabia

Mohd Muzzammil Hussain
MNR Dental College
NTR Health University
Vijayawada, India

Waseem Mohammed Abdul
Department of Biological Sciences
King Abdulaziz University
Jeddah, Saudi Arabia

Syed Shoeb Iqbal Razvi
Department of Biochemistry
King Abdulaziz University
Jeddah, Saudi Arabia

ISSN 2192-3698   ISSN 2192-3701 (electronic)
SpringerBriefs in Public Health
ISBN 978-3-030-04335-3   ISBN 978-3-030-04336-0 (eBook)
https://doi.org/10.1007/978-3-030-04336-0

Library of Congress Control Number: 2018962494

This Springer imprint is published by the registered company Springer Nature Switzerland AG
The registered company address is: Gewerbestrasse 11, 6330 Cham, Switzerland

*This book is dedicated to **Abu al-Qasim Khalaf ibn al-Abbas Al-Zahrawi** (commonly known as Albucasis). He was born at Madinat al-Zahra near Cordoba in Islamic Spain (936–1013 CE). He is considered as the greatest surgeon in the Islamic medical tradition. His comprehensive medical texts, combining Middle Eastern and Greco-Roman classical teachings, shaped European surgical procedures up until the Renaissance.*

His greatest contribution to history is Kitab al-Tasrif, a 30-volume collection of medical practice, of which large portions were translated into Latin and in other European languages.
(Image and Info. Source: http://www.muslimheritage.com/article/abu-al-qasim-al-zahrawi-great-surgeon)

Preface

Oral health loss is one of the major problems existing all over the world. Oral cavity is home to numerous pathogenic microorganisms, some of which are responsible for the progression and development of various systemic diseases, such as cancer, diabetes, and myocardial infections. Worldwide, 60–90% of school children and nearly 100% of adults have dental cavities, often leading to pain and discomfort. Oral health loss has severe implications on global economy and human health before, during, and after the onset of any of the oral diseases. Economically developing countries are facing a financial crunch due to the continuous burden of investing resources for the upgradation of healthcare systems in maintaining oral hygiene. About 70% of oral cancers are preceded by onset of precancerous oral lesions. There is an urgent need to identify some natural and economic solutions to treat the oral diseases, which can be affordable by each individual.

The present book covers the issues related to oral health, oral diseases, and the role of medicinal plants in overcoming the oral health issues. Negligence of oral hygiene is a major drawback in inviting numerous oral pathogens and in turn making oral cavity susceptible to many life-threatening diseases. We have also highlighted the research gaps in dealing with the oral health-related problems all over the world.

We are thankful to Professor Syed Safiullah Ghori. Associate Professor, Anwar ul Uloom Collge of Pharmacy, Hyderabad, Telangana, India for his suggestions and support while finalizing this book.

Jeddah, Saudi Arabia   Khalid Rehman Hakeem
Jeddah, Saudi Arabia   Waseem Mohammed Abdul
Vijayawada, India   Mohd Muzzammil Hussain
Jeddah, Saudi Arabia   Syed Shoeb Iqbal Razvi

Contents

Chapter 1
Introduction

Medicinal plants have been used as traditional remedies in numerous human diseases for centuries around the globe (Dias et al. 2012). In some rural parts of the developing countries, the first hand source of medicine is the local traditional medicine, due to their low cost and previous success rates (Sofowora et al. 2013). The importance of traditional medicine has increased to an extent that about 80% of the people in developing countries use traditional medicines for their health care in one or the other way (Ekor 2013). The medicinal plant-based natural products have been reported to have a unique collection of bioactive compounds; the present medicinal plant-based drugs are the evidence of new entities used in pharmaceutical industries (Yuan et al. 2016). In terms of a diversity of diseases caused by various microorganisms, the biggest problem nowadays is the increasing resistance by these microbes against a wide variety of currently used therapeutic agents, and this has renewed the interest and need to discover the novel anti-infective compounds (Nathan 2012). According to the previous reports, there are around 500,000 species of plants occurring throughout the world, among which only about 1% plants has been tested for drug discovery profile, still there is great potential for exploiting the novel leads from the plant species (Cragg and Newman 2013). Many studies have been carried out in last few decades to evaluate the efficacy of medicinal plants and traditional medicine in treatment of various oral disorders. Many plant-derived medicines used in traditional medicinal systems have been recorded in pharmacopeias as agents used to treat infections and a number of these have been recently investigated for their efficacy against oral microbial pathogens (Palombo 2011). The basic antimicrobial activities pertaining to traditional medicine and other medicinal plant products including essential oils have been reviewed previously (Swamy et al. 2016). Therefore, the purpose of this book is to display some useful examples from the literature review that have been reported to validate the traditional use of numerous medicinal plants against numerous oral health deformities. On a broad perspective, the research focusing on traditional and medicinal treatment strategies used to cure or inhibit numerous oral health-degrading organisms is presented. The activity of this traditional medicine is in many dimensions such as reduction of dental plaque development, influencing the bacterial adhesion to surfaces and reduction of the symptoms of oral diseases will be discussed subsequently. In addition to this, there are few clinical studies that have investigated the safety and efficacy of such plant-derived medicines will be described. Despite the widespread use of different sources of fluoride, dental caries continues to be the single most prevalent and costly oral infectious disease worldwide. Virulent biofilms that are tightly adherent to oral surfaces are a primary cause of infectious diseases in the mouth, including dental caries (Jeon et al. 2011). Dental caries are a result of interactions of the things, which include specific bacteria with their metabolic/virulence products, salivary constituents, and other dietary carbohydrates that are present on susceptible tooth surface (Kilian et al. 2016). Dental caries pathogenesis is modulated by various virulence factors such as the acidification of the milieu, formation of the extracellular polysaccharide (EPS)-rich biofilm matrix, and the maintenance of a low-pH environment at the interface of tooth-biofilm (Koo et al. 2013). Biofilms formed *in vivo* consists of *Streptococcus mutans* as the primary producers of the EPS-rich matrix although it is flooded with mixed flora. Due to its presence as a primary occupant, *Streptococcus mutan*s is responsible for virulent biofilms development although there are few reports stating the involvement of other microbes also in pathogenesis of the disease (Kim et al. 2018). This mechanism of pathogenesis involves a three-step process before turning into a cariogenic bacterium. The factors include effective utilization of dietary sucrose for the rapid synthesis of EPS by the activity of glucosyl transferase (Gtfs) and a fructosyl transferase, secondly the adhesion to glucan-coated surfaces, and finally the acidogenic and acid-tolerant nature. By this mechanism, the *S. mutans* exposes out from the complex oral microbiome and modulates effectively from nonpathogenic to cariogenic biofilms (Koo et al. 2013).

Bacterial plaque plays a major role in the development of dental biofilm also known as oral biofilm; it is a sticky, colorless film that forms on teeth/dental prosthesis consistently, and this feature is ubiquitous to human beings only. It comprises 50 bacterial species with thousands of bacteria, and it needs an extracellular matrix permitting the maturation of microorganism and its aggregation. Oral biofilm reappears after a long period of its removal by brushing the tooth. The major causes among various other reasons of caries and periodontal disease is the consistent display of bacterial plaque between the teeth. The existence of bacterial plaque in teeth/prostheses is a serious issue, pertaining to various oral diseases such as osteomyelitis, candidiasis, and peri-implantitis among others. In recent few years, numerous studies have been carried out regarding the connection between the oral biofilm presence and consequent gingival inflammation with many systemic diseases through three major mechanisms: the hematogenous dissemination of oral biofilm bacteria, the effect of infection on adjacent tissues and cavities, and/or inflammatory mechanisms. There is a marked scientific evidence about periodontal disease as a risk factor for dementia, cardiovascular disease, and diabetes. Apart from this, there are reports by some researchers that periodontal disease is also associated with chronic obstructive pulmonary disease, low newborn weight, the metabolic syndrome, and rheumatoid arthritis, among others. Oral hygiene (teeth, prostheses, and soft oral tissues) comprises of the primary preventive measure against various oral diseases and the associated systemic diseases, such as aspiration pneumonia. Due to lack of awareness and education in some rural areas and elderly populations, the mental and physical capacities are limited, which plays a major role in chemical control (mouthwashing) and brushing. According to reports, there is a great decline in oral health status among the elder population, specifically in people with severe cognitive or functional impairments, which is responsible for making them compromise on their capability to maintain oral hygiene practices without any assistance (Razak et al. 2014). Thus, in a study carried out among the middle-aged adults by Naorungroj and co-workers reported that decrease in two cognitive measures were slightly associated with lack of proper tooth brushing only and not with other conditions such as clinical gingivitis, dental behaviors, or with periodontitis. Likewise, a study of the institutionalized elderly by Steinmassl et al. observed that tooth hygiene indices were irresponsible with the degree of cognitive impairment as evaluated by the Shulman clock-drawing test. Hence, further research is needed in a broad perspective where the loss of functional or cognitive capacities alters the maintenance of acceptable oral health levels and on the factors related with a possible decrease in oral hygiene practices. The above data supports the progress and implementation of programs responsible for plaque control prevention for both the dependent and independent old age people, especially in those with suspected or diagnosed cognitive impairment or dementia.

Chapter 2
Oral Hygiene for Healthy Life

Oral health is essential and integral to general health; it is a determinant factor for quality of life as both are associated strongly. Healthy primary and permanent teeth plays a major role in maintenance of health and well-being during the journey of life. There is a need for healthy and well-functioning dentition during every style of life. Since it is associated and responsible for supporting essential body functions of a human such as eating, speaking, socializing, and smiling (Sischo and Broder 2011). The individual shape and form to the face is given by the teeth. Thus, maintenance of oral hygiene is an essential part of life (Razak et al. 2014).

Oral hygiene is the most crucial component of our overall general health. Oral health contributes to a healthy lifestyle (Lasemi et al. 2014). Almost all the pathogenic microorganisms tend to enter the human body through mouth. Most of the bacteria and other microorganisms are generally destroyed in the mouth due to the salivary antibacterial action (Takeshita et al. 2016). However, when the contamination is more and the microorganisms are in high number, the consumed food is spoiled then multiple unfavorable implications are observed on the general health (Balto et al. 2016). Maintenance of oral health is the key to healthy life. Any kind of malfunctioning in the oral cavity leads to disease condition which may be direct or indirect (Kumar et al. 2016). How well we care for our teeth is directly proportional to the overall general health of the body.

There are many adverse effects of deteriorating oral health on human health; some of them are even responsible for the cause of death (Kane 2017). Any kind of negligence in maintenance of oral health is making a way towards decline in healthy life. The impact of oral health loss on human life every day is pervasive and subtle, influencing sleep, rest, eating, and social factors (Lago et al. 2017). The complete process of health deterioration due to oral hygiene loss is gradual, and this is the major factor for ignorance of any diseases condition until it starts getting out of the hand (Singh et al. 2013).

During the life course, oral tissues and teeth are generally exposed to numerous environmental factors, which are directly or indirectly responsible for causing

SpringerBriefs in Public Health, https://doi.org/10.1007/978-3-030-04336-0_2

diseases or even tooth. The most common oral diseases include tooth decay and periodontal diseases loss (Li et al. 2000). The effect of oral diseases may be direct to a particular area of the human body but their impact and consequences affect the whole body (Kilian et al. 2016). According to WHO, oral health is defined as a state of being free from mouth and facial pain, oral infection, throat and oral cancer, periodontal diseases, tooth loss, and tooth decay. Furthermore, the other diseases and disorders can diminish an individual's capacity in chewing, speaking, biting, and psychological wellness.

The combination of healthy mouth and a healthy body are interrelated for maintenance of oral hygiene in contrast (Bhat and Srinivasan 2018). There are detrimental consequences on psychological and physical well-being due to poor oral health. Therefore, the huge burden of oral diseases represents a strongly underestimated public health challenge for almost every country worldwide (Settineri et al. 2017).

Oral diseases are mostly invisible and hidden or they are not given much importance until and unless there is some extremity. Due to this, they are accepted as an unavoidable consequence of ageing and life. However, there is clear evidence that oral cavity-related health loss is not inevitable, but can be prevented or decreased by some simple and effective ways at all stages of life, both at population and at individual level (Griffin et al. 2012). Oral hygiene is diminished by several pathogenic microbes dwelling in the oral region by various damages, which includes dental cavities, a trauma from injuries, oral maladies, periodontal (gum) diseases, and oral cancer. As the mouth is considered as a mirror of the body, reflecting symptoms of systemic diseases are few at times (Sanz et al. 2013). Prevention and control of oral pathogens is an important factor for maintenance of good oral hygiene (Par et al. 2014). There are various ways of preventing the bacterial growth in the oral cavity but once dwelling of oral microbiota is initialized, their presence may be responsible for numerous complications ultimately leading to oral health loss (Dagli et al. 2016).

Oral health deformities are mostly neglected and in majority of countries, specifically developing countries, there has been subsequent failure to formulate oral health policies. Policies of both general health and oral health need to be formulated based on life course perspective, with a focus on both social and temporal.

Chapter 3
Oral Diseases and Their Severity

Oral diseases are among the topmost public health issue globally with significant socioeconomic impacts and on the other hand, they are neglected most of the times in the public health policy (Preet 2013). The data mined about oral health from Global Burden of Disease Study in 2010 by Marcenes et al. revealed that periodontal disease, oral cancer, edentulism, and palate collectively accounted for about 18,814,000 disability-adjusted life years. The global burden of these diseases augmented by an average of 45.6% from the year 1990 to 2010 in parallel with the major non-communicable diseases such as diabetes 69% (Marcenes et al. 2013). Both non-communicable diseases and oral diseases are closely interlinked by sharing common risk factors which include excess sugar consumption, use of tobacco, and finally the most important factor is the mechanism, i.e., infection/inflammatory pathways (Jin et al. 2015). The long-term sustainable approach on promotion of health and prevention will be successful with effective and collaborative efforts. Oral health and general health are interrelated with each other.

3.1 Dental Caries: Widespread Chronic Disease

3.1.1 Cardiovascular Diseases (CVD)

Cardiovascular disease is one of the most common chronic diseases affecting significant proportion of population in developed countries leading to mortalities and morbidities (Nag and Ghosh 2013). According to a report from World Health Organisation, almost 31% of all deaths are due to CVD globally every year, CVD was reported to be the commonest cause in case of death globally in the year 2013, accounting for about 17 million of total 54 million deaths (WHO, Report 2017). The most common causes leading to CVD include hypertension, smoking, tobacco, family genetic background, diabetes, limited physical activity, obesity, unhealthy

© The Author(s), under exclusive license to Springer Nature Switzerland AG 2019
K. R. Hakeem et al., *Oral Health and Herbal Medicine*,
SpringerBriefs in Public Health, https://doi.org/10.1007/978-3-030-04336-0_3

eating habits, and lifestyle (AHA 2015). Furthermore, periodontal disease has been recognized as an emerging and potential risk factor of CVD (Nazir 2017). Moreover, a decade ago, several studies demonstrated that the total loss of teeth and periodontal disease had high risk of developing CVD. Holmlund and colleagues investigated if there is any relation between the number of remaining teeth and severity of periodontal disease is related to any past history of hypertension (HT) and heart attack (Holmlund et al. 2006). It was reported that the severity of the periodontal disease was directly related to the HT, independent of the age factor and the prevalence of heart attack in the population of middle-age group, While the number of teeth was related to the prevalence of the heart attack independent of the age. Altogether, this study showed that the periodontal disease and loss of teeth from any cause are closely associated with the development of CVD. Additionally, the same authors performed a bigger study later with 7674 subjects and concluded that there is a linear relationship between loss of teeth and CVD (Holmlund et al. 2010). Nevertheless, the number of teeth remaining can be used to measure the risk of CVD in adults. While in another study, Bahekar et al. mentioned that both the existing and new coronary heart cases significantly increased in periodontal disease (Bahekar et al. 2007).

3.1.2   Diabetes

Most of the diabetic patients are unaware of the oral complications that arise from the disease and the obscure reality that the oral health and diabetes are related to each other. In a random survey conducted by a research group on 500 diabetic patients, it was concluded that only 28% of the patients asserted that they monitor their periodontal health regularly by periodic visits to dentist. While 48% were conscious of the increased susceptibility and oral health complications and 38% recognized that periodontal health can affect blood sugar levels (Noble et al. 2009).

There is considerable evidence in the literature that the patients with poorly controlled diabetes have higher prevalence of development of periapical lesions (Bender and Bender 2003; Segura-Egea et al. 2012). A recent clinical study has presented a significant association between the incidences of periapical lesions and endodontic treatments in DM2 patients. Moreover, a study published in 2011 with regard to the rate of endodontic treatments reported that there was lower success rate in the primary root canal treatments compared to the healthy individuals; however, both the groups showed similar success rate with secondary root canal treatment (López-López et al. 2011). In fact, the diabetic patients are more susceptible to infections leading to loss in the dental health with severe periodontal disease. Therefore, periodontal disease is considered as the "Sixth complication" of diabetes (Negrato and Tarzia 2010). Nevertheless, poorly controlled diabetics are more susceptible to periodontal diseases than well-controlled diabetics (Bender and Bender 2003). Emerging studies suggest that the periodontal diseases predict end-stage kidney disease in diabetics. A study published in the year 2006 showed worse long-term control of blood sugar and cardiovascular complications in diabetics with periodontal disease as compared to the healthy individuals (Jansson et al. 2006). Many health care providers neglect the fact that there is a two-way relationship between the diabetes and periodontal disease as evidenced by many recent studies. A study published recently evaluated the awareness of 232 physicians and 278 dentists to relate the diabetes and tooth loss due to periodontal diseases (Jansson et al. 2006). It was rather surprising to observe that only 50% of the participants believed that diabetic patients with periodontal disease are susceptible to tooth loss compared with non-diabetic individuals (Jansson et al. 2006). However, the authors concluded that there needs to be increased awareness to relate diabetes and periodontal diseases among dentists and physicians to effectively prevent and manage tooth loss and its associated complications. Indeed, a multifaceted approach is the need of the hour to combat periodontal diseases in the form of journals, periodicals, case reports, and specific clinical guidelines for dentists and physicians. Taylor et al. insists that a relatively small and concerted effort can go a long way in reducing the burden of diabetes on patients, families, and societies (Taylor et al. 2010). Furthermore, undiagnosed diabetes can be detected early by periodic oral health examination. Li et al. have developed clinical guidelines for dentists, which could be of immense help in diagnosing diabetes in early stages, which could reduce the burden of health care and recurring expenditures in treating periodontal diseases (Li et al. 2011).

3.1.3   Respiratory Disease

Apart from CVD and diabetes, respiratory (lung) disease is also a major cause of death in developed countries. A first ever study was performed few years ago which studied the association of oral health with the deaths due to CVD, cancer, and respiratory diseases in older Japanese subjects. After a Aichi Gerontological Evaluation Study (AGES) and data obtained from 4425 respondents through a questionnaire, it was concluded that the deaths due to CVD and respiratory diseases were predictable (not cancer deaths) by oral health in older Japanese patients (Azarpazhooh and Leake 2006). Moreover, a systematic review was performed in 2006, which studied the relation between respiratory diseases and oral health. It was reported that there was a considerable evidence that the professional maintenance of oral health reduced respiratory diseases in older patients in nursing homes and intensive care units (ICU). Also, there was significant evidence that the risk of pneumonia and oral health were interdependent (Azarpazhooh and Leake 2006). However, it was also mentioned that there was poor evidence regarding the association between chronic obstructive pulmonary disease (COPD) and oral health. However, one study reported that the prevalence of periodontitis was nearly 44% in COPD patients compared to the control group (7.3%) which was significant after adjusting age, gender, number of cigarettes smoked (Leuckfeld et al. 2008). However, to arrive a concrete conclusion, large-scale studies are required.

3.1.4   Stroke

The commonest kind of strokes are cerebrovascular ischemic strokes which result from a clot in the blood vessel supplying blood to brain, hardening of arteries due to deposition of fat in the lining of blood vessels could be the underlying cause. Stroke has been reported as the third major cause of death in the countries of developed world after deaths due to cancer and heart disease. After heart disease, the stroke has been asserted because of the hardening of arteries. A study performed by Beck and colleagues implicated the role of oral health in strokes in USveterans. These patients are known to have higher strokes; however, the results due to poor oral health could have been underestimated, as the authors did not separate the data of hemorrhagic strokes from ischemic strokes. However, Morrison et al. later reported a non-significant rise in the strokes due to periodontal health. In a different study by Wu and colleagues, the relation between the periodontal health and fatal and non-fatal stroke was studied. It was concluded that there was a 17% rise in the risk of stroke in patients with severe periodontitis compared to the healthy individuals (Wu et al. 2000). However, the dentists prescribe antibiotic medication for the patients with severe periodontal disease and heart disease to prevent the leakage of bacteria into the bloodstream, which will otherwise damage valves leading to subacute bacterial endocarditis (SBE). Therefore, this is again directly proportional to the severity of infection and inflammation (Pallasch and Slots 1996).

3.1.5   Chronic Kidney Disease

The oral health in chronic kidney disease patients is often overlooked due to lack of awareness about the consequences of the poor periodontal health. This further leads to severe health complications like hardening of arteries, protein-energy wasting, and inflammation resulting in deaths. Nonetheless, the complications of poor oral health are more significant in patients with severe CKD, aged patients with concurrent medications, diabetes, and reduced immunity leading to multiple ailments. However, the role of improved oral health in reducing the incidences of CKD is yet to be fully understood. In a recent published review, Meurman et al. have mentioned that the risk of peripheral vascular disease (PVD) is increased in patients with poorly maintained oral health. The reduced blood flow to the organs, for example, legs due to PVD is a consequence of hardening of arteries, which again could be related to the poor oral health. However, this need to be further clarified by supporting studies (Akar et al. 2011).

3.1.6   Dementia

It is interesting to note that there exists a relation between poor dental health and Alzheimer's disease and dementia (Gatz et al. 2006; Stein et al. 2007; Stewart and Hirani 2007). In a distinguished study performed on 4000 Japanese subjects aged

65 years or older and who underwent dental treatments and psychiatric assessment showed that the participants with fewer or no natural teeth experienced memory loss compared to healthy individuals (Okamoto et al. 2010). Moreover, impaired memory loss and forgetfulness has been associated to periodontal disease as evidenced by blood tests (Noble et al. 2009).

3.1.7   Stomach Ulcers

It has been reported the poor oral hygiene is the underlying cause of stomach ulcers by *Helicobacter pylori*. This is supported by the discovery of *Helicobacter pylori* in both plaques in teeth and stomach. Hence, it has been professed that the teeth could be the reservoir for *H. pylori*, which could be a potential source of transmission and persistent source of infection (Al Asqah et al. 2009).

3.1.8   Oral Cancers

There has been increased incidences of oral cancers recently due to excessive use and abuse of alcohol and tobacco. Moreover, use of mouthwashes containing alcohol and periodontal disease has been implicated as independent causes in the development of head, neck, and esophageal cancers (Guha et al. 2007). Researchers have been busy in establishing a link between the periodontal disease and oral cancers during 1990s (Beck et al. 1996). In severe cases, the brushing of teeth and chewing may often release the harmful bacteria in the bloodstream, which may further lead to complications like endotoxemia and bacteremia resulting in an overall rise in the inflammatory mediators like interleukin-6, C-reactive protein, and fibrinogen (Moura da Silva et al. 2012; Spahr et al. 2006). For instance, many studies have identified harmful, periodontal disease-causing bacteria in the patients with hardened arteries in heart and elsewhere such as amniotic fluid (Zi et al. 2014). Hence, it is essential to prevent the periodontal diseases, which could further reduce the incidences of oral cancers.

3.2   Different Oral Health Deformities

Oral health is destroyed in many ways and thus resulting in various diseased conditions. Many tooth deformities are interrelated and are gradually leading to tooth loss or any kind of oral health loss. The following are some of the diseases related to oral cavity:

3.2.1   Caries

This is a condition where the dissolution of the organic substance takes place resulting in multifactorial etiology and ultimately the demineralization of the inorganic
part of the tooth takes place. This process of dentine demineralization and enamel is
due to organic acids that are deposited in dental plaque by microbial activity, and by
the process of anaerobic metabolism of sugars taken by diet (Bang and Kristoffersen
1972; Sicca et al. 2016).

3.2.2   Hypoplasia

Enamel hypoplasia results from incomplete or defective formation of the enamel
organic matrix, usually associated with genetic or environmental factors. It is a dis-
order caused by a dysfunction in enamel matrix secretion during the mineralization
or maturation of this tissue. When the cause of this condition is hereditary, the
enamel malformations may come from defects in the genes that encode the proteins
related to the mineralization process. Thus, when that happens, there is involvement
of both the primary and secondary dentitions in a generalized way (Carvalho et al.
2013).

3.2.3   Dental Erosion

"Dental erosion is a unique form of teeth deterioration which eventually occurs
without involvement of bacteria but due to the chemical corrosion of extrinsic and
intrinsic acids. Dental erosion (erosive tooth wear) is the situation of a chronic loss
of dental hard tissue that is chemically etched away from the tooth surface by acid
and/or chelation without bacterial involvement. Acids of intrinsic (gastrointestinal)
and extrinsic (dietary and environmental) origins are the main etiologic factors.
Rampant caries is defined as quickly spreading caries that are affecting at least two
of the upper incisors. In epidemiologic studies, rampant caries is defined as a
decayed, missed, and filled teeth (DMFT) value of 5 or more, and labial caries is
regarded as a specific entity (Cheng et al. 2009).

3.2.4   Periodontal Disease

"The pathology of periodontal disease emerges from gum and reaches the periodontal ligament right up to the alveolar bone." The most important risk factor in the
development of periodontal disease is represented by inadequate oral hygiene along

with improper diet. The focus of global periodontal epidemiology during the last half century has been on identifying populations who have periodontal disease and situations, where disparities in disease prevalence exist between groups. Unlike dental public health activities directed towards dental caries, less effort has been made in periodontal epidemiology with regard to surveying or monitoring groups who may be at greater risk for moderate or severe disease and evaluating public health initiatives directed at mitigating risk or reducing such periodontal disease prevalence.

3.2.5 Potentially Malignant Oral Lesions

These are the pathologies of the oral mucous area (leukoplakia, oral lichen planus) that represent the tendency for continuous degeneration of oral health if some favorable conditions persist (Mortazavi et al. 2014) (Figs. 3.1 and 3.2).

3.3 Reasons of Declined Oral Health

There are various reasons by which the oral health is affected, and many factors are responsible for decline of oral health (Gambhir and Gupta 2016). Some common factors, which are primarily responsible for deteriorating oral health include use of tobacco, poor dietary practices, unhygienic conditions, and negligence of initial dental problems (Erik et al. 2003). In this aspect, there is a strong interrelationship

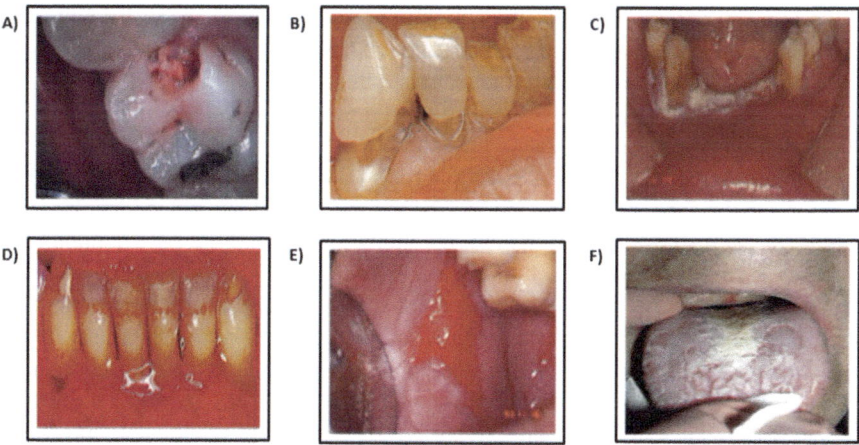

Fig. 3.1 (**a**) Caries of the teeth (**b**) Hypoplasia and pits on the surface of the enamel (**c**) Dental Erosion (**d**) Periodontal disease (**e**) *Oral lichen planus* (**f**) Oral leukoplakia (Reproduced from Sicca et al. 2016)

Fig. 3.2 (**a, b**) Oral cancer
(Reproduced from Sicca
et al. 2016)

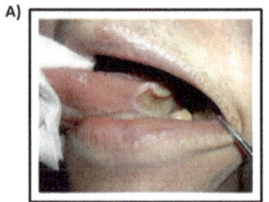 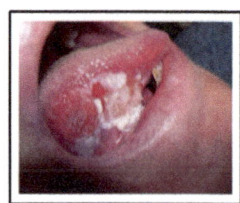

between oral diseases and non-communicable chronic diseases (Türp and Spranger 2016). Factors such as diet, lifestyle, and socioeconomic status are mainly responsible for oral health loss (Patrick et al. 2006). Oral health is usually linked with general health in older people due to various their inability to resist multiple infections and microbial attack.

Our mouths are filled with large amount of microorganisms (Tada and Hanada 2010). Bacteria in the oral cavity utilize the sugar from food and convert them into acids (Gupta et al. 2013). As the time passes, the acids attack the tooth there by creating tooth decay and subsequently leading to a cavity (Featherstone 2008). Furthermore, these bacteria along with mucus and some other particles form a sticky and colorless film on teeth known as "plaque." Flossing and proper brushing can help to get rid of plaque. However, when this is ignored and the plaque is hardened, it forms "tartar," which cannot be removed by brushing. At this stage, scaling by a dental professional can remove this tartar. The longer the plaque and tartar on teeth, the more harmful they are. After this stage, the advancement of deposits will be responsible for removal of teeth (extraction). The bacteria are responsible for inflammation of gums, a condition known as "gingivitis." This is a condition where the gums turn red, swollen, and bleeding is also frequent. Although, the gum becomes susceptible due to the mild gum disease, i.e., gingivitis, there is a strong chance of reversal and retaining normality when there is constant brushing, flossing, and scaling by dentist. However, there is no bone and tissue loss in this form of gum disease. Furthermore, untreated gum disease can develop into a condition known as "periodontitis" (inflammation around teeth). At this stage, the gums are pulled away from the teeth leading to formation of spaces known as "pockets" that are infected. As the infection spreads, the body's immune system defends the bacteria when they are growing below the gum line. Bacterial toxins and the response of body starts to break down the connective tissue and the bone that are holding the teeth in place. At this stage, if this decay is not treated, the gums, bones, and the tissue supporting teeth are destroyed (Yadav and Prakash 2016).

The dentistry plays a major role for the diagnosis of oral health loss associated with diet. Reliable nutrition strategies are important for the improvement of general health. Negligence in proper diet is markedly linked with augmented probabilities of diseases related to oral cavity (Gupta et al. 2015). The advice of proper diet plays a major role in preventing the onset of various diseases, both general and oral (Dale et al. 2014). Inadequate training of professionals in providing nutritional information is directly responsible for inconsistencies in dietary advice. Previous studies have reported that the awareness about nutrition to dentists and dietitians (Reed 2014).

Diet effects the growth of the oral cavity: Based on nutritional imbalance stages, which may be early, or late nutritional imbalance, there are different consequences. Malformations are mostly related to nutritional imbalance (Reynolds et al. 2015). Moreover, there is a marked transformation in periods of intense growth with periods of relative quiescence in different components of the stomatognathic apparatus: it is clear that a great damage can be expected when there is nutritional imbalance in a very active growth period (Singh et al. 2011). In fact, shortage of minerals and other supplements during the phase of pre-conception effects the development of embryo, manipulating dental organogenesis, the skull/facial development, and growth of the maxilla (Belcastro et al. 2007; Dion et al. 2007).

Chapter 4
Traditional Information About Herbal Medicine of Oral Activity

Since prehistoric times, human have been using various natural products such as microbes, plants, animals, and marine organisms to treat and alleviate diseases (Yuan et al. 2016). Traditional therapists are well thought out to be among the effective representatives of transformation as they have a strong hold in command authority among their communities, these people include legal and political advisors, marriage and family counselors, and physicians (Rudolph et al. 2007). There are numerous success stories of traditional medicine in maintenance of general health and oral health by various entities. Among the major advantages of traditional medicine is the expertise of the traditional healer and the cases handled by them, the use of medicines according to the old scripts mentioning the importance of a particular medicine. Various natural products such as Chinese traditional medicine, Ayurveda, Kampo, Unani, traditional Korean medicine and Homeopathy etc. These types of treatments have been practiced since thousands of years in different parts of the world and blossomed into orderly regulated systems of medicine (Anushri et al. 2015; Karygianni et al. 2016).

Currently, the use of natural products and traditional medicines have tremendous and incomparable advantages, which includes abundant clinical experiences, biological activities, and their unique diversity of chemical structures (Amirkia 2016).

Natural products evolved over millions of years ago have a unique diversity in their chemical structure responsible for their drug-like properties and biological activity (Dias et al. 2012). Other than natural products, there are some traditional methods used to treat diseases in terms of treatment of oral cavity-related treatment; there is a method known as "oil pulling" (Naseem et al. 2017). Oil pulling is proved to maintain and improve oral health. For the first time the importance of oil pulling came into light by Dr. F karach (Parolia 2009; Peedikayil et al. 2015). The application of oil pulling was explored from its knowledge of Ayurveda. This treatment is of two types: first is Gandoosha, in which the complete mouth is filled with oil and gargling is impossible. The second is Kavala Graha, where lesser amounts of oil is used so that gargling is possible (Peedikayil 2015; Sooryavanshi and Mardikar 1994).

Organic oils such as sesame oil, coconut oil, and sunflower oil are useful particularly if it is cold pressed although refined oil can also be used for pulling the bacteria, protozoa, and viruses from oral cavity (Parolia 2009). Olive oil contains 70% monounsaturated fatty acids with major composition of oleic acid. It also consists of phenolic compounds such as phytosterols, squalin, vitamin A, E, and K. These constituents are having antioxidative, antimicrobial, and immunomodulatory effects (Kensche et al. 2013). Mouth rinse containing almond oil is found to decrease gingival scores and olive oil-based mouth rinses inhibits plaque formation (Bekeleski et al. 2012). Coconut oil contains monolaurin, which is found to be effective against microorganisms such as *Helicobacter pylori, Staphylococcus aureus, Enterobacter* spp., and *Candida* spp. Furthermore, monolaurin also has virucidal activity by dissolving lipids and phospholipids in the viral envelope, ultimately leading to virus disintegration (Thaweboon et al. 2011). In a study by Anand et al., there was 20% reduction in bacterial number upon 40 days of oil pulling using sesame oil. Apart from this, there was reduction in the severity of dental caries. It was presumed that bacteria and toxins from the body may be removed through tongue and are trapped in the oil and subsequently thrown out from the body (Anand et al. 2008).

A randomized pilot trial with 20 adolescent subjects has revealed that oil pulling with sesame oil is as effective as chlorhexidine for reduction in microbes associated with it and halitosis (Asokan et al. 2011).

Oil pulling is found to be effective for maintenance of oral health when practiced regularly and correctly. However, oil pulling is not recommended by American dental association, and it does not replace dental therapy completely (Oklahoma Dental Association Patient's page 2014).

Numerous traditional medicinal plants have been evaluated and assessed for their activity and use in the treatment or prevention of oral diseases. Hamza et al. reported Tanzanian plants, due to the bioactive components present in them, are responsible for treatment of oral diseases (Hamza et al. 2006). Different plant parts are used for treatment of numerous oral diseases. For instance, *Isodon rugosus* dried leaves are considered as a good remedy for toothache. Decoction of leaves from *Olea europaea* is used as gargle to combat mouth and gum diseases (Zougagh 2018).

Chapter 5
Role of Medicinal Plant Species in Oral Health Sector

The current global need for some alternative treatment and prevention options for oral diseases that are effective, safe, and economical is due to increase in incidence of diseases, emergence of antimicrobial resistance bacteria, opportunistic infections in immune compromised persons, and financial burden (Khan et al. 2017). Although many agents are being commercially used for treatment of oral microbiota, their undesirable side effects make them less successful in safety aspects (Fair and Tor 2014). Some of the chemical antibacterial agents such as cetylpyridinium chloride, chlorhexidine, and amine fluorides have shown to exhibit some kind of toxicity with staining of teeth, leading to oral cancer. Therefore, the search for alternative substances is in great demand and bioactive compounds from plants are being used in traditional medicines as complementary medicine (Chandra shekar et al. 2016; Palombo 2011).

Medicinal herbs are gaining importance and herbal renaissance is observed all over the world. The herbs and their products are symbolizing safety in contrast to the synthetic medicine, which are considered as unsafe to both environment and humans (Karunamoorthi et al. 2013; Karimi et al. 2015).

Herbs comprising medicinal properties are a valuable and effective source for treatment of various diseases (Petrovska 2012). These herbal extracts have been consistently used in maintaining oral health by tooth cleaning and as antimicrobial plaque agents (Balto et al. 2017).

5.1 Combating Oral Diseases with Herbal Medicine

The herbal medicine usage is successfully increasing around the globe. The herbal medicine has gained a marked momentum in various national health care settings (Yuan et al. 2016). The use of herbal extracts in dentistry is due to various advantages such as antimicrobial plaque agents, reducing inflammation, antiseptics,

K. R. Hakeem et al., *Oral Health and Herbal Medicine*,
SpringerBriefs in Public Health, https://doi.org/10.1007/978-3-030-04336-0_5

antioxidants, antifungals, antivirals, and analgesic. Furthermore, the herbal medicine is effective in controlling microbial plaque in gingivitis, wound healing, and periodontitis (Cruz Martínez et al. 2017).

The application of herbal remedies presumed an international interest, which has culminated in their being used in combating various diseases and ailments in both developing and well-developed countries (Sinha and Sinha 2014). Furthermore, currently only a few herb-based drugs are approved for their admirable medicinal properties, a vast majority of naturally occurring medicinal herbs are considered only as food supplements due to lack of studies about clinical trials (Parveen et al. 2015). However, in recent few years much importance was given in exploring different herbs in dentistry.

Various medicinal herbs are having applications in maintenance of oral hygiene by suppressing various oral microbes and by other curative ways (Sravani et al. 2015).

5.2   Medicinal Plants Used in Dentistry

5.2.1   Aloe vera *(Fig. 5.1a)*

Aloe vera comprises the chemical constituents such as saccharides, anthraquinones, fatty acids, and prostaglandins. Some other substances include vitamins, minerals, enzymes, amino acids, gibberellin, cholesterol, uric acid, lignins, triglycerides, steroids, salicylic acid, and beta-sitosterol. It is analgesic, antiviral, antifungal, antioxidant immune modulating, antibacterial, antiseptic, and anti-inflammatory. *Aloe vera* is used at the sites of periodontal surgery, aphthous ulcers, toothpick injuries, lichen

Fig. 5.1 (**a**) *Aloe vera* (**b**) *Sanguinaria canadensis* (**c**) *Vaccinium macrocarpon* (**d**) *Matricaria recutita* (**e**) *Rhizoma cimicifugae* (**f**) *Carum carvi*

planus, chemical burns, dry socket, gum abscesses, gingival problems associated with leukemia and AIDS, migratory benign pemphigus, glossitis, geographic tongue and burning mouth syndrome, candidiasis, desquamative gingivitis, acute mono-cytic leukemia denture sore mouth, vesiculobullous diseases, and xerostomia. Studies have reported that it might lead to allergic reactions: popular dermatitis and generalized eczematous (Taheri et al. 2011; Wynn 2005; WHO Report 1999).

5.2.2   *Blood Root (*Sanguinaria canadensis*) (Fig. 5.1b)*

The major constituent of *Sanguinaria canadensis* is sanguinarine, which have the medicinal properties such as anti-inflammatory, antibacterial, and antifungal prop-erties. This plant is mainly used for remineralization of enamel lesions, acute sore throat, gingivitis, and periodontal disease. However, in case of children and preg-nant or lactating women it is considered as unsafe. It has some side effects when used for long term such as glaucoma, stomach pain, diarrhea, edema, heart disease, nausea and vomiting, miscarriage, visual changes, and paralysis (Taheri et al. 2011).

5.2.3   *Cranberry (*Vaccinium macrocarpon*) (Fig. 5.1c)*

Cranberry has many medicinal properties and the major constituents include poly-phenols and flavonoids, which have been reported to have anticarcinogenic, antibac-terial, antiviral, antifungal, and antioxidant properties. Because of its antiadhesive property, it is used against periodontal disease, dental caries, and oral squamous cell carcinoma. Furthermore, there are no reports of adverse effects (Oswal and Charantimath 2011; Kukreja and Dodwad 2012; Yoo et al. 2011).

5.2.4   *Chamomile (*Matricaria recutita*) (Fig. 5.1d)*

The chemical composition of chamomile consists of essential oils, volatile oils, and chamazulene. Other constituents include α-bisabolol, flavonoids, luteolin, and related sesquiterpenes, quercetin, and apigenin. Presence of above active ingredi-ents is responsible for its antibacterial and antiviral activity, antispasmodic, anti-inflammatory, and smooth muscle-relaxing action. Major uses include in gingivitis, periodontal disease also in ulcers as a mouthwash. Generally, chamomile is consid-ered safe during pregnancy or breast-feeding. However, it is not recommended for the people with allergies to plants of the Asteraceae family (aster, ragweed, and chrysanthemums), and mugwort pollen (Sudarshan and Vijayabala 2012; Kamat et al. 2011; Taheri et al. 2011).

5.2.5   Black Cohosh (Rhizoma Cimicifugae racemosae) (Fig. 5.1e)

The principle constituents of black cohosh are acetylacetone, cycloartenol-based triterpenes action, 26 deoxy acetol, 26-deoxyactein, cimidenol, and cimicifugaside. It has been reported to be an anti-inflammatory property. This feature is used for treatment of periodontitis although there is not much evidence and studies about it. It is contraindicated during pregnancy and lactation and in children under the age of 12 years. Minor adverse effects of black cohosh include headache and gastrointestinal upset (Taheri et al. 2011).

5.2.6   Caraway (Carum carvi) (Fig. 5.1f)

The major components of caraway are carvone (50–60%) and limonene (40%). It also contains 3–7% volatile oil; the medicinal properties of caraway are antihistaminic, expectorant, antiseptic, antimicrobial, anti-inflammatory, spasmolytic, and flavoring agent (Mardani et al. 2015).

5.3   Medicinal Plants Used to Maintain Oral Health

5.3.1   Evening Primrose (Oleum oenothera biennis) (Fig. 5.2a)

Chemical constituents of primrose include g-linolenic acid, linoleic acid (cis-linoleic acid) which are present more than 60%, followed by oleic acid about 10%, (cis-g-linolenic acid) above 10%, stearic acid and palmitic acid less than 10%. *Oleum oenothera biennis* has shown antiallergic activity, antiulcer activity. It is used in treatment of dental caries and orthodontic tooth movement. Some rare side effects include headaches, nausea, and diarrhea (Wiesner 2017; Matsumoto-Nakano et al. 2011).

5.3.2   Garlic (Allium sativum) (Fig. 5.2b)

The presence of components such as diallyl sulfide, alliin, S-acetylcysteine, ajoene, dithiin, vitamins B, enzymes, proteins, and minerals. It has antiviral, antibacterial, bacteriostatic, antifungal, antiseptic, and antihelminthic effects. Garlic was tested for treatment of periodontitis and dental caries, and few reports have demonstrated adverse effects such as asthmatic attacks, increased bacterial attachment to

Fig. 5.2 (**a**) *Oleum oenothera* (**b**) *Allium sativum* (**c**) *Zingiber offcinalis* (**d**) *Commiphora myrrha* (**e**) *Camellia sinensis* (**f**) *Azadirachta indica*

orthodontic wires, and contact dermatitis (Mohammad et al. 2014; Oswal and Charantimath 2011; Kamat et al. 2011).

5.3.3   Ginger (Zingiber officinalis) (Fig. 5.2c)

The multiple components of ginger include oleoresin, 1–4% essential oils, zingiberene, curcumin, bisabolene, and sesquiphellandrene along with alcohol and monoterpene aldehydes. The medicinal properties include antibacterial, anti-inflammatory, and analgesic property. The use of ginger is also reported for relieving toothache and for the treatment of oral thrush. Furthermore, ginger may reduce the toxic effects of the cyclophosphamide, a chemotherapeutic agent. Use of ginger is not preferred during pregnancy and patients with the biliary disease. Due to interference of ginger with blood clotting, a special care should be taken in patients undergoing treatment on anticoagulant therapies such as coumadin or heparin (Sudarshan and Vijayabala 2012; Azizi et al. 2015).

5.3.4   Myrrh (Commiphora myrrha) (Fig. 5.2d)

There are three major constituents of myrrh, which include the volatile oil, the resin, and the gum. The gum consists of 65% carbohydrates, 20% proteins, and is composed of 4-*O*-methylglucuronic acid, arabinose, and galactose. Myrrh has various applications, both in general health and oral health maintenance such as in gingivitis, pharyngitis, ulcers, tonsillitis, and stomatitis. Topical application is used for the treatment of infections of the oral region. Nevertheless, it should be avoided during pregnancy. Some side effects include contact dermatitis (Oswal and Charantimath 2011; Al-Mobeeriek 2011).

5.3.5   Green Tea (Camellia sinensis) (Fig. 5.2e)

The polyphenol contents in green tea comprises catechin (C), gallocatechin (GC), epicatechin gallate (ECG) epicatechin (EC), epigallocatechin (EGC), and epigallo-catechin gallate. It is antibacterial, anti-inflammatory, and antiviral. *Camellia sinensis* is reported to be used in the treatment of periodontal disease (Florêncio Passos et al. 2018; Kukreja and Dodwad 2012; Wolfram 2007; Corwin 2009; Sultan et al. 2016; Naauman et al. 2017).

5.3.6   Neem (Azadirachta indica) (Fig. 5.2f)

Neem is rich with many useful constituents such as sodium nimbinate, azadi-rachtin, salannin, nimbin, nimbidin, genin, nimbidiol, and quercetin. Leaves of neem consist of carbohydrates, fiber, and about ten amino acid proteins, carot-enoids, calcium, and fluoride. Neem has a wide range of activities such as anti-microbial, antitumor, analgesic, antibacterial, antiviral, antifungal, anti-inflammatory, antihelminthic, anticariogenic, antipyretic, and antioxidant activity. There are some reports about the neem and its components, which are used in the treatment of gingivitis, dental caries, and periodontitis. External applications: 70% ethanol extract of the leaves is diluted to 40%. This diluted extract must be applied (Kukreja and Dodwad 2012; Dhingra and Vandana 2016; Bodiba et al. 2018).

5.4   Medicinal Plants Used in the Treatment of Oral Diseases

5.4.1   Thyme (Thymus vulgaris) (Fig. 5.3a)

The main constituents of thyme are phenols, carvacrol, and thymol. A salve obtained from thyme, goldenseal, and myrrh is used for treatment of oral herpes. Furthermore, there are reports of treatment of halitosis and chronic candidiasis by using thyme. It must be carefully used in young children, pregnant and lactating mothers. However, there are some side effects such as vomiting, dizziness, and breathing difficulties. Some people are sensitive towards thyme oil when applied on the skin or used as a mouth rinse (Kukreja and Dodwad 2012; Taheri et al. 2011).

Fig. 5.3 (**a**) *Thymus vulgaris* (**b**) *Curcuma longa* (**c**) *Ocimum sanctum* (**d, e**) *Salvadora persica* (**f**) *Mentha piperita*

5.4.2   Turmeric *(*Curcuma longa*) (Fig. 5.3b)*

Curcuma longa contains a variety of bioactive components, which include a number of monoterpenes and sesquiterpenes such as zingiberene, β-turmerone, and curcumin Alpha. The coloring is due to curcuminoids, around 60% among them are a mixture of monodesmethoxy, curcumin and bis-desmethoxy curcumin. Various pharmacological properties of turmeric include anticarcinogenic, antibacterial, antimutagenic, and antioxidant used in treatment of dental caries, gingivitis, halitosis, pit and fissure sealant, and oral lichen planus. There is a great relief observed when the aching teeth was massaged with finely grinded and roasted turmeric powder which reduces the swelling also (Chaturvedi 2009; Nagpal and Sood 2013).

5.4.3   Tulsi *(*Ocimum sanctum*) (Fig. 5.3c)*

Tulsi is a traditional medicinal plant, which consists of eugenol, tannins, and few essential oils. It has also got few bioactive components such as methyl chavicol, linalool, and 1,8-cineole. It is antiulcer, antimicrobial, antihelminthic, analgesic, and antipyretic, and is used in the treatment for periodontal diseases (Kukreja and Dodwad 2012).

5.4.4   Meswak *(*Salvadora persica*) (Fig. 5.3d, e)*

Chewing sticks have been widely used in Africa, Indian subcontinent, and the Middle East since ancient times. Meswak is a derivative obtained from Arak tree and is used by people of different cultures and in many developing nations as a traditional toothbrush to maintain oral hygiene (Al lafi et al. 1995). This is a cheap and easily affordable natural toothbrush suitable for cleansing teeth, possesses various medicinal properties, and is easily available in developing countries. The Meswak extract has also found to be effective in the dentifrices as antiplaque and antigingivitis agents (Gupta et al. 2012). Chewing sticks should be obtained from fresh stems of medicinal plants. It is believed that chewing on these stems facilitate salivary secretions which possibly help in oral cleaning and control of plaque.

5.4.5   *Peppermint (*Mentha piperita*) (Fig. 5.3f)*

Peppermint leaves are having some chemical constituents, which include 0.1–1.0% volatile oil, which comprises of menthone (20–31%) and menthol (29–48%). The oil is having muscle-relaxing action and is analgesic. One traditional use of peppermint oil is its application for toothache just by soaking a cotton ball in the oil and rubbing it on the tooth or placing it in the cavity. Peppermint oil is to be avoided by people with inflammation of the gallbladder, severe liver damage, or obstruction of bile ducts. Some marked side effects include headache, perianal burning, bradycardia, burning and gastrointestinal upset, skin rashes, muscle tremors, heartburn, and ataxia. Some in vivo studies are also being performed (Tardugno et al. 2017; Taheri et al. 2011; Gupta et al. 2017).

5.5   Medicinal Plants in Nourishing Oral Health

5.5.1   *Sesame (*Sesamum indicum*) (Fig. 5.4a)*

This sesame plant (*Sesamum indicum*) is a precious gift to humankind from nature due to its varied health effects and remarkable nutritional qualities. Oil pulling is a traditional practice in ayurvedic medicine involving swishing of oil in the mouth for both oral and systemic health benefits (Hebbar et al. 2010). The oil from sesame seed is used mostly due to its desirable health benefits and varied medicinal properties. When compared to chlorhexidine, oil pulling therapy with sesame oil has numerous advantages such as no prolonged after taste, no staining, and no allergy. Sesame oil is more cost-effective when compared to chlorhexidine and is easily available as a grocery item in most households (Asokan et al. 2010).

Fig. 5.4 (**a**) *Sesamum indicum* (**b**) *Lavandula angustifolia* (**c**) *Melaleuca alternifolia* (**d**) *Salvia offcinalis*

5.5.2   Lavender Oil *(Lavandula angustifolia) (Fig. 5.4b)*

The oil is extracted from the *Lavandula angustifolia* flowers. There are reports of stress reduction, decrease in anxiety, and improvement in the mood when inhaled or orally administered but at high anxiety levels does not favor the positive effects of the oil. It is used to reduce patients' anxiety in dental clinics. It is reported to be an anxiolytic agent, particularly when it is used in waiting area. Furthermore, it is also used during surgical procedures, as it has been demonstrated to decrease the pain due to needle insertion (Lehrner et al. 2005; Kim et al. 2011).

5.5.3   Tea Tree Oil *(Melaleuca alternifolia) (Fig. 5.4c)*

Melaleuca alternifolia is a native of Australia with antifungal and antiseptic properties with a mild solvent (Arweiler et al. 2000). The products of this plant are used in the treatment of throat irritation, wounds, burns, stings, and skin infections of all kinds. The method of using it is rubbing tree tea oil directly on inflamed gum and sore for temporary relief. The tree tea mouthwash is used for soothing oral inflammation. There is a little solvent action observed and hence has wide and potential applications in root canal treatment for necrotic pulp tissue dissolution. Mouthwashes with tea tree oil have proved to be effective for oral candidiasis patients (Filoche et al. 2005).

5.5.4   Sage *(Salvia officinalis) (Fig. 5.4d)*

Sage volatile oil consists of major constituents as alpha and beta-thujone, cineole, and camphor. Apart from it, there are other bioactive components such as tannins, rosmarinic acid, and flavonoids. Other than this, the use of sage is prevalent in treatment of sore throat, inflammations, and gingivitis. Sage oil is reported to have various medicinal properties including antifungal, antibacterial, and antiviral, to be avoided by children, when there is high fever and in pregnant women. However, fewer side effects such as increased heart beat and mental disturbance is observed in some cases. High dose may cause convulsions (Taheri et al. 2011; Beheshti-Rouy et al. 2015).

Chapter 6
Oral Health Care Products Obtained from Medicinal Plants

Abbreviations

ANZCTR	Australian New Zealand Clinical Trials Registry
CCTR	Chinese Clinical Trial Registry
CPPCT	Cuban Public Registry of Clinical Trials
CRIS-K	Clinical Research Information Service of Republic of Korea
CTRI	Clinical Trials Registry of India
EU-CTR	European Union Clinical Trials Register
GCTR	German Clinical Trials Register
ICMJE	International Committee of Medical Journal Editors
IRCT	Iranian Registry of Clinical Trials
ISRCTN	The International Standard Randomized Controlled Trial Number
JPPN	Japan Primary Registries Network
NCT	United States Trial Registry—ClinicalTrials.gov
NNTR	The Netherlands National Trial Register
PACTR	Pan African Clinical Trial Registry
R&D	Research & Development
RCTHI	Randomized Clinical Trials of Herbal Interventions
ReBEC	Brazilian Clinical Trials Registry
SLCTR	Sri Lanka Clinical Trials Registry
TCTR	Thai Clinical Trials Register

Despite a large number of studies are being carried out for testing the biological effects of natural products (NP) and their phytochemical constituents, only a little part among them are able to attain the clinical phase and are commercially available now. In this view, an analysis was performed by Freires et al. to study about the medicinal plant research impact on oral health care for the last 15 years. All the plants and their details related to clinical trials and their mode of action are stated in Table 6.1.

Abbreviations given above are the various clinical trial registries around the globe.

K. R. Hakeem et al., *Oral Health and Herbal Medicine*,
SpringerBriefs in Public Health, https://doi.org/10.1007/978-3-030-04336-0_6

Table 6.1 Different health care products obtained from medicinal herbs

Registry	Sponsor	Product specification	Active ingredient(s)	Dental condition	Outcomes
NCT, 2014	Tatyasaheb kore dental college	Mouthwash (Khadir chaal)	Acacia catechu extract	Gingivitis	Changes in gingival index
NCT, 2014	Johnson & Johnson	Mouthwash (Listerine)	Menthol, methyl salicylate, eucalyptol, and thymol	Plaque formation and gingivitis	Changes in mouth modified gingival index (MGI)
NCT, 2014	University of Santiago	Mouthwash (Listerine)	Menthol, eucalyptol, and methyl salicylate	Gingivitis and plaque formation	Reduction in bacterial viability and biofilm thickness
NCT, 2015	University hospital Ghent and Johnson & Johnson	Mouthwash (Listerine)	Menthol, thymol, eucalyptol, and methyl salicylate	Chronic periodontitis	Microbiological and clinical changes
CTRI, 2015	GLAXOSMITHKLINE ASIA Pvt. Ltd. and colgate palmolive	Toothpaste (Parodontax)	*Salvia officinalis*, mentha, limonene	Gingivitis	Reduction in plaque and gingival inflammation
EU-CTR, 2006	Laboratories expanoscience, France	Oral capsule (Piasdedine 300)	Avocado (Persea gratissima) fruit oil, and soybean (Glycine max) seeds	Chronic periodontitis	Changes in gingival inflammation index
NCT, 2013	Tatyasaheb kore dental college	Septin tablet	Rubia cordifolia, eEmblica officinalis, and tinospora cordifolia	Chronic periodontitis	Reduction in serum C-reactive protein levels, pocket depth, and clinical attachment level
NCT, 2013	University of Santiago and Johnson & Johnson	Mouthwash (Listerine)	Thymol, menthol	Chronic periodontitis and plaque formation	Reduction in bacterial viability (%), biofilm thickness (microns), and covering grade (%)

(continued)

Table 6.1 (continued)

Registry	Sponsor	Product specification	Active ingredient(s)	Dental condition	Outcomes
NCT, 2013	University of Taubate	Mouthwash (Listerine total care)	Menthol, thymol, eucalyptolm and methyl salicylate	Gingivitis and plaque formation	Changes in plaque and gingival indices

In the present world, due to marked research and success in the field of natural product (NP)-based drugs or formulations a large number of conditions responsible for deteriorating oral health can be prevented, ameliorated, and/or treated. There are a huge number of NP-inspired drugs and NP-derived drugs, toothpastes, mouthwashes, etc. that have been available under prescription or over the counter (Cragg and Newman 2016). There are a wide range of reasons due to which the discovery of natural product-based drugs is still under process: short- and long-term toxicity, microbial resistance, high costs for the end user, adverse and side effects, and compromised sustainability of industrial large-scale production are among many others. Thus, there is a great need for the discovery of more potent, low-cost, safe, effective, and well-tolerated drugs and oral care formulations in dentistry.

Chapter 7
Clinical Evidence of Dental Treatment by Using Herbal Formulations

The use of herbal formulations has gained momentum in recent past with many people using natural product-based substances to stay away from dental health issues. In the present study, the patient is treated by using herbal mouthwash containing *Baccharis dracunculifolia*. Figure 7.1 displays the result of treatment accompanied by herbal mouthwash. This study was carried out under the NTR Health University.

K. R. Hakeem et al., *Oral Health and Herbal Medicine*,
SpringerBriefs in Public Health, https://doi.org/10.1007/978-3-030-04336-0_7

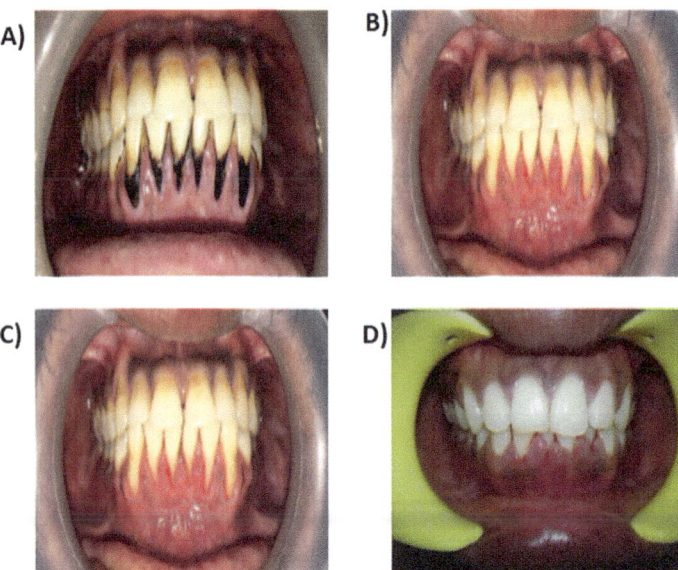

Fig. 7.1 Use of herbal formulation in cleansing of teeth (before and after treatment) (**a** and **c**) before treatment, (**b** and **d**) after treatment

Chapter 8
Conclusions

After reviewing the case reports and previously reported documented literature described in this book, it can be concluded that further research is needed to include the active herbal extracts or phytochemicals in already existing treatment regimens as supportive therapy or an entirely new regimen. Moreover, it is also crucial to assess the safety and efficacy of the plant extracts, purified phytochemicals, and essential oils in the ongoing clinical trials that can help to reduce the overall burden of oral diseases worldwide. More importantly, specific research projects must be designed and executed concerning the issues of adequate population size with suitable statistical power, quality control, and standardization of purified compounds or extracts.

The prevailing oral health problems like dental caries, periodontitis, microbial plaque, and gingivitis could be prevented and cured if suitable herbal remedies are used consistently over a period. The dental professionals should also emphasize on these products in their prescriptions and recommendations as these are natural, safe, and economic. The common complaints reported in many cases like loss of enamel and demineralization of enamel could be avoided using herbal interventional therapy.

Finally, the regular dental checkups and increase in the frequency of cleansing the teeth is quite important in maintaining oral health and avoiding systemic diseases in the end. People have the mindset of visiting the dentist only when the pain becomes quite unbearable and cannot be controlled by medicines; hence, regular checkups ensure that the dental health is maintained properly to avoid complications in future. We believe that the nipping infections in the bud could extend the healthy dental years and lessen the overall health care costs of any economy.

K. R. Hakeem et al., *Oral Health and Herbal Medicine*,
SpringerBriefs in Public Health, https://doi.org/10.1007/978-3-030-04336-0_8

Chapter 9
Future Recommendations

9.1 Role of Dental Professionals in Improving Safety

Dental professionals have an important role in terms of educating common people and creating awareness regarding the drawbacks and risks of maintaining poor oral health. There should be a separate treatment chart for each patient, including the ongoing medicine for any other illness of the particular patient. Some of the non-steroidal anti-inflammatory drugs and aspirin cause bleeding with dental treatment. Various factors are to be taken into consideration before dental surgery as the herbs altering the liver function have been reported to alter the metabolism of drugs used in dentistry.

Awareness about safe usage along with training, collaboration and communication among providers of medicines, particularly traditional medicine is of utmost importance. Ayurveda uses various metals in therapeutics, but it is only after due purification process. Until the end of nineteenth century, there was a steep increase in traditional medicine usage initializing from a home level to industrial production. There are numerous registered pharmaceutical industries around the world, which utilizes the traditional medicine either directly or by modifying its basic structure; finally, the use of herbal medicines is being increasing rapidly due to its safety efficacy.

On the other hand, there is a high demand to retain the safety of people along with effective quality control of the herbal preparations. There is a great need of research to determine and conceptualized loss of oral health as a major contributor for total health destruction.

Research recommendations focusing on economic, psychological, and social impacts of oral deformities and treatment:

- Sensitivity testing of basic health indicators for individuals suffering from various oral disorders and other conditions.
- Considering the sickness illness profile modification for use in patients with various oral conditions.
- Establishing the relationships among clinical symptoms of disease and subjective indicators assessing disease impact.
- Evaluating measures and indicators in population of all ages.

References

Akar H, Akar GC, Carrero JJ, Stenvinkel P, Lindholm B (2011) Systemic consequences of poor oral health in chronic kidney disease patients. Clin J Am Soc Nephrol 6(1):218–226

Al Asqah M, Al Hamoudi N, Anil S, Al Jebreen A, Al-Hamoudi WK (2009) Is the presence of Helicobacter pylori in dental plaque of patients with chronic periodontitis a risk factor for gastric infection? Can J Gastroenterol 23(3):177–179

Al-Mobeeriek A (2011) Effects of myrrh on intra-oral mucosal wounds compared with tetracy-cline- and chlorhexidine-based mouthwashes. Clin Cosmet Investig Dent 3:53–58

American Heart Association (2015) What is cardiovascular disease

Amirkia V (2016) Plant extracts and natural products—predictive structural and biodiversity-based analyses of uses, bioactivity, and 'research and development' potential. http://discovery.ucl.ac.uk/1527357/1/20161101%20Amirkia%20Thesis.pdf

Anand TD, Pothiraj C, Gopinath RM, Kayalvizhi B (2008) Effect of oil-pulling on dental caries causing bacteria. Afr J Microbiol Res 2:63–66

Anushri M, Yashoda R, Puranik MP (2015) Herbs: a good alternatives to current treatments for Oral health problems. Int J Adv Health Sci 1(12):26–32

Arweiler NB, Donos N, Netuschil L, Reich E, Sculean A (2000) Clinical and antibacterial effect of tea tree oil—a pilot study. Clin Oral Investig 4:70–73

Asokan S, Emmadi P, Chamundeswari R (2009) Effect of oil pulling on plaque induced gingivitis: a randomized, controlled, triple-blind study. Indian J Dent Res 20:47–51

Asokan S, Kumar RS, Emmadi P, Raghuraman R, Sivakumar N (2011) Effect of oil pulling on halitosis and microorganisms causing halitosis: a randomized controlled pilot trial. J Indian Soc Pedod Prev Dent 29:90e94

Azarpazhooh A, Leake JL (2006) Systematic review of the association between respiratory dis-eases and oral health. J Periodontol 77(9):1465–1482

Azizi A, Aghayan S, Zaker S, Shakeri M, Entezari N, Lawaf S (2015) In vitro effect of *Zingiber officinale* extract on growth of *Streptococcus mutans* and *Streptococcus sanguinis*. Int J Dent 2015:1–5

Bahekar AA, Singh S, Saha S, Molnar J, Arora R (2007) The prevalence and incidence of cor-onary heart disease is significantly increased in periodontitis: a meta-analysis. Am Heart J 154(5):830–837

Balto H, Al-Sanie I, Al-Beshri S, Aldrees A (2017) Effectiveness of Salvadora persicaextracts against common oral pathogens. Saudi Dent J 29(1):1–6

Balto H, Al-Sanie I, Al-Beshri S, Aldrees A (2017) Effectiveness of Salvadora persica extracts against common oral pathogens. The Saudi Dental Journal 29(1):1–6

K. R. Hakeem et al., *Oral Health and Herbal Medicine*,
SpringerBriefs in Public Health, https://doi.org/10.1007/978-3-030-04336-0

Bang G, Kristoffersen T (1972) Dental caries and diet in an Alaskan Eskimo population. Scand J Dent Res 80(5):440–444

Beck J, Garcia R, Heiss G, Vokonas PS, Offenbacher S (1996) Periodontal disease and cardiovascular disease. J Periodontol 67(10s):1123–1137

Beheshti-Rouy M, Azarsina M, Rezaie-Soufi L, Alikhani MY, Roshanaie G, Komaki S (2015) The antibacterial effect of sage extract (Salvia officinalis) mouthwash against Streptococcus mutans in dental plaque: a randomized clinical trial. Iran J Microbiol 7:173–177

Bekeleski GM, McCombs G, Melvin WL (2012) Oil pulling: an ancient practice for a modern time. J Int Oral Health 4:1–10

Belcastro G, Rastelli E, Mariotti V, Consiglio C, Facchini F, Bonfiglioli B (2007) Continuity or discontinuity of the life-style in central Italy during the Roman imperial age-early middle ages transition: diet, health, and behavior. Am J Phys Anthropol 132(3):381–394

Bender IB, Bender AB (2003) Diabetes mellitus and the dental pulp. J Endod 29:383–389

Benjamin EJ, Virani SS, Callaway CW et al (2018) Heart disease and stroke statistics-2018 update: a report from the American Heart Association. Circulation 137:e67–e492

Bhat SS, Srinivasan G (2018) Healthy mouth and healthy body. In: AIIMS conference proceedings. https://doi.org/10.13140/RG.2.2.10844.49286

Bodiba DC, Prasad P, Srivastava A, Crampton B, Lall NS (2018) Antibacterial activity of Azadirachta indica, Pongamia pinnata, Psidium guajava, and Mangifera indica and their mechanism of action against Streptococcus mutans. Pharmacogn Mag 14(53):76–80

Carvalho LD, Bernardon JK, Bruzi G, Andrada MA, Vieira LC (2013) Hypoplastic enamel treatment in permanent anterior teeth of a child. Oper Dent 38(4):363–368

Chandra Shekar BR, Nagarajappa R, Singh R, Suma S, Thakur R (2016) Antimicrobial efficacy of the nilotica, Murrayakoenigii (Linn.) Sprengel, Eucalyptus, and Psidium guajava on primary plaque combinations of Acacia colonizers: an in vitro study. Indian J Dent Res 27:415–420

Chaturvedi TP (2009) Uses of turmeric in dentistry: an update. Indian J Dent Res 20:107–109

Cheng R, Yang H, Shao M, Hu T, Zhou X (2009) Dental erosion and severe tooth decay related to soft drinks: a case report and literature review. J Zhejiang Univ Sci B 10(5):395–399

Chicago Dental Society (2011) Good oral health starts with exercise, eating right. CDS Rev 104(2):34

Corwin A (2009) Herbal supplements: healthcare implications and considerations. Can Dent Hyg Assoc 24:7–15

Cragg GM, Newman DJ (2013) Natural products: a continuing source of novel drug leads. Biochim Biophys Acta 1830(6):3670–3695

Cruz Martínez C, Diaz Gómez M, Oh MS (2017) Use of traditional herbal medicine as an alternative in dental treatment in Mexican dentistry: a review. Pharm Biol 55:1992–1998

Dagli N, Dagli R, Darwish S, Baroudi K (2016) Oral microbial shift: factors affecting the microbiome and prevention of oral disease. J Contemp Dent Pract 7(1):90–96

Dale J, Lindenmeyer A, Lynch E, Sutcliffe P (2014) Oral health: a neglected area of routine diabetes care? Br J Gen Pract 64(619):103–104

DebMandal M, Mandal S (2011) Coconut (Cocos nucifera L.: Arecaceae): in health promotion and disease prevention. Asian Pac J Trop Med 4:241–247

Dhingra K, Vandana KL (2016) Effectiveness of Azadirachta indica (neem) mouthrinse in plaque and gingivitis control a systematic review. Int J Dent Hyg 15:4–15

Dias DA, Urban S, Roessner U (2012) A historical overview of natural products in drug discovery. Metabolites 2(2):303–336

Dion N, Cotart JL, Rabilloud M (2007) Correction of nutrition test errors for more accurate quantification of the link between dental health and malnutrition. Nutrition 23(4):301–307

Dye BA (2012) Global periodontal disease epidemiology. Periodontol 2000 58:10–25

Dye BA, Barker LK, Li X, Lewis BG, Beltrán-Aguilar ED (2011) Overview and quality assurance for the oral health component of the National Health and Nutrition Examination Survey (NHANES). J Public Health Dent 71(1):54–61

Ekor M (2013) The growing use of herbal medicines: issues relating to adverse reactions and chal-
    lenges in monitoring safety. Front Pharmacol 4:177–187
Erik Petersen P (2003) Tobacco and Oral health – the role of the World Health Organization. In:
    Oral health & preventive dentistry, vol 1, pp 309–315
Fair RJ, Tor Y (2014) Antibiotics and bacterial resistance in the 21st century. Perspect Medicin
    Chem 6:25–64
Featherstone J (2008) Dental caries: a dynamic disease process. Aust Dent J 53:286–291
Filoche SK, Soma K, Sissons CH (2005) Antimicrobial effects of essential oils in combination
    with chlorhexidine digluconate. Oral Microbiol Immunol 20:221–225
Gambhir R, Gupta T (2016) Need for oral health policy in India. Ann Med Health Sci Res
    6(1):50–55
Gatz M, Mortimer JA, Fratiglioni L, Johansson B, Berg S, Reynolds CA, Pedersen NL (2006)
    Potentially modifiable risk factors for dementia in identical twins. Alzheimers Dement
    2:110–117
Gordon N (2007) Oral health care for children attending a malnutrition clinic in South Africa. Int
    J Dent Hyg 5(3):180–186
Griffin SO, Jones JA, Brunson D, Griffin PM, Bailey WD (2012) Burden of oral disease among
    older adults and implications for public health priorities. Am J Public Health 102(3):411–418
Guha N, Boffetta P, Wünsch Filho V, ElufNeto J, Shangina O, Zaridze D, Curado MP, Koifman S,
    Matos E, Menezes A, Szeszenia Dabrowska N, Fernandez L, Mates D, Daudt AW, Lissowska
    J, Dikshit R, Brennan P (2007) Oral health and risk of squamous cell carcinoma of the head
    and neck and esophagus: results of two multicentric case-control studies. Am J Epidemiol
    166(10):1159–1173
Gupta P, Agarwal N, Anup N, Manujunath BC, Bhalla A (2012) Evaluating the anti-plaque efficacy
    of meswak (*Salvadora persica*) containing dentifrice: a triple blind controlled trial. J Pharm
    Bioallied Sci 4:282–285
Gupta P, Gupta N, Pawar AP, Birajdar SS, Natt AS, Singh HP (2013) Role of sugar and sugar sub-
    stitutes in dental caries: a review. ISRN Dent 2013:519421
Gupta M, Gupta M, Abhishek (2015) Oral conditions in renal disorders and treatment consider-
    ations—a review for pediatric dentist. Saudi Dent J 27(3):113–119
Gupta AK, Mishra R, Singh AK, Srivastava A, Lal RK (2017) Genetic variability and correlations
    of essential oil yield with agro-economic traits in *Mentha* species and identification of promis-
    ing cultivars. Ind Crop Prod 95:726–732
Hamza OJM, van den Bout-van den Beukel CJP, Matee MIN, Moshi MJ, Mikx FHM, Selemani
    HO, Mbwambo ZH, Van der Ven AJAM, Verweij PE (2006) Antifungal activity of some
    Tanzanian plants used traditionally for the treatment of fungal infections. J Ethnopharmacol
    108(1):124–1324
Hebbar A, Keluskar V, Shetti A (2010) Oil pulling-unraveling the path tomysticure. J Int Oral
    Health 2:11–14
Holmlund A, Holm G, Lind L (2006) Severity of periodontal disease and number of remaining
    teeth are related to the prevalence of myocardial infarction and hypertension in a study based
    on 4,254 subjects. J Periodontol 77(7):1173–1178
Holmlund A, Holm G, Lind L (2010) Number of teeth as a predictor of cardiovascular mortality in
    a cohort of 7,674 subjects followed for 12 years. J Periodontol 81(6):870–876
Jansson H, Lindholm E, Lindh C, Groop L, Bratthall G (2006) Type 2 diabetes and risk for peri-
    odontal disease: a role for dental health awareness. J Clin Periodontol 33(6):408–414
Jeon JG, Rosalen PL, Falsetta ML, Koo H (2011) Natural products in caries research: current (lim-
    ited) knowledge, challenges and future perspective. Caries Res 45(3):243–263
Jéssica L, Laiza F, Andréia M, Ligia P (2017) Educational program in oral health for caregivers on
    the oral hygiene of dependent elders. Rev Odontol 46:284–291
Jin LJ, Lamster IB, Greenspan JS, Pitts NB, Scully C, Warnakulasuriya S (2015) Global burden
    of oral diseases: emerging concepts, management and interplay with systemic health. Oral Dis
    22(7):609–619

Joshipura KJ, Hung HC, Rimm EB, Willett WC, Ascherio A (2003) Periodontal disease, tooth loss, and incidence of ischemic stroke. Stroke 34:47–52

Kamat S, Rajeev K, Saraf P (2011) Role of herbs in endodontics: an update. Endodontology 23:98–102

Kane SF (2017) The effects of oral health on systemic health. Gen Dent 65:30–34

Karimi A, Majlesi M, Rafieian-Kopaei M (2015) Herbal versus synthetic drugs; beliefs and facts. J Nephropharmacol 4(1):27–30

Karunamoorthi K, Kaliyaperumal J, Jegajeevanram V, Embialle B (2013) Traditional medicinal plants: a source of Phytotherapeutic modality in resource-constrained health care settings. Evid Based Complement Alternat Med 18:67–74

Karygianni L, Al-Ahmad A, Argyropoulou A, Hellwig E, Anderson AC, Skaltsounis AL (2016) Natural antimicrobials and oral microorganisms: a systematic review on herbal interventions for the eradication of multispecies oral biofilms. Front Microbiol 16(1529):1–17

Kensche A, Reich M, Kümmerer K, Hannig M, Hannig C (2013) Lipids in preventive dentistry. Clin Oral Investig 17:669–685

Khan HA, Baig FK, Mehboob R (2017) Nosocomial infections: epidemiology, prevention, control and surveillance. Asian Pac J Trop Biomed 7(5):478–482

Kilian M, Chapple IL, Hannig M et al (2016) The oral microbiome—an update for oral healthcare professionals. Br Dent J 221:657–666

Kim S, Kim HJ, Yeo JS, Hong SJ, Lee JM, Jeon Y (2011) The effect of lavender oil on stress, bispectral index values, and needle insertion pain in volunteers. J Altern Complement Med 17:823–826

Kim D, Liu Y, Raphael IB, Hiram S, Áurea S-S, Yong L, Geelsu H, Micha F, David RA, Hyun K (2018) Bacterial-derived exopolysaccharides enhance antifungal drug tolerance in a cross-kingdom oral biofilm. ISME J. https://doi.org/10.1038/s41396-018-0113-1

Koo H, Falsetta ML, Klein MI (2013) The exopolysaccharide matrix: a virulence determinant of cariogenic biofilm. J Dent Res 92:1065–1073. https://doi.org/10.1177/0022034513504218

Kukreja BJ, Dodwad V (2012) Herbal mouthwashes: a gift of nature. Int J Pharma Bio Sci 3:46–52

Kumar NN, Panchaksharappa MG, Annigeri RG (2016) Psychosomatic disorders: an overview for oral physician. J Indian Acad Oral Med Radiol 28:24–29

Lago JD, Fais LMG, Montandon AAB, Pinelli LAP (2017, Sept-Oct) Educational program in oral health for caregivers on the oral hygiene of dependent elders. Rev Odontol UNESP 46(5):284–291

Lassemi E, Motamedi MHK, Frouzandeh A, Sarkarat F, Ghasemi M et al (2014) Histopathologic changes in dental follicles of bone-impacted vs. partially bone-impacted 3rd molars. J Oral Hyg Health 2:120. https://doi.org/10.4172/2332-0702.1000120

Lehrner J, Marwinski G, Lehr S, Johren P, Deecke L (2005) Ambient odors of orange and lavender reduce anxiety and improve mood in a dental office. Physiol Behav 86:92–95

Leuckfeld I, Obregon-Whittle MV, Lund MB, Geiran O, Bjørtuft Ø, Olsen I (2008) Severe chronic obstructive pulmonary disease: association with marginal bone loss in periodontitis. Respir Med 102(4):488–494

Lewis HA, Rudolph MJ, Mistry M, Monyatsi V, Marambana T, Ramela P (2004) Oral health knowledge and original practices of African traditional healers in Zonkizizwe and Dube, South Africa. SADJ 59:243–246

Li X, Kolltveit KM, Tronstad L, Olsen I (2000) Systemic diseases caused by oral infection. Clin Microbiol Rev 13(4):547–558

Li S, Williams PL, Douglass CW (2011) Development of a clinical guideline to predict undiagnosed diabetes in dental patients. J Am Dent Assoc 142(1):28–37

López-López J, Jané-Salas E, Estrugo-Devesa A, Velasco-Ortega E, Martín-González J, Segura-Egea JJ (2011) Periapical and endodontic status of type 2 diabetic patients in Catalonia, Spain: a cross-sectional study. J Endod 37:598–601

Lugan R, Niogret MF, Leport L, Guegan JP, Larher FR, Savoure A, Kopka J, Bouchereau A (2010) Metabolome and water homeostasis analysis of *Thellungiella salsuginea* suggests that dehydration tolerance is a key response to osmotic stress in this halophyte. Plant J 64:215–229

Lupi-Pégurier L, Muller-Bolla M, Fontas E, Ortonne JP (2007) Reduced salivary flow induced by systemic isotretinoin may lead to dental decay. A prospective clinical study. Dermatology 214(3):221–226

Marcenes W, Kassebaum NJ, Bernabé E, Flaxman A, Naghavi M, Lopez A, Murray CJ (2013) Global burden of oral conditions in 1990-2010: a systematic analysis. J Dent Res 92(7):592–597

Mardani M, Afra SM, Tanideh N, Tadbir AA, Modarresi F, Koohi-Hosseinabadi O, Iraji A, Sepehrimanesh M (2015) Hydroalcoholic extract of Carum carvi L. in oral mucositis: a clinical trial in male golden hamsters. Oral Dis 22:39–45

Matsumoto-Nakano M, Nagayama K, Kitagori H, Fujita K, Inagaki S, Takashima Y, Tamesada M, Kawabata S, Ooshima T (2011) Inhibitory effects of Oenothera biennis (Evening Primrose) seed extract on streptococcus mutans and S. mutans-induced dental caries in rats. Caries Res 45:56–63

Mohammad SG, Raheel SA, Baroudi K (2014) Clinical and radiographic evaluation of *Allium sativum* oil as a new medicament for vital pulp treatment of primary teeth. J Int Oral Health 6(6):32–36

Morrison HI, Ellison LF, Taylor GW (1999) Periodontal disease and risk of fatal coronary heart and cerebrovascular diseases. J Cardiovasc Risk 6:7–11

Mortazavi H, Baharvand M, Mehdipour M (2014) Oral potentially malignant disorders: an overview of more than 20 entities. J Dent Res Dent Clin Dent Prospects 8(1):6–14

Moura da Silva G, Coutinho SB, Piscoya MD, Ximenes RA, Jamelli SR (2012) Periodontitis as a risk factor for pre-eclampsia. J Periodontol 83(11):1388–1396

Naauman Z, Maliha S, Yousaf A, Anas A, Usman Z, Mohammad A (2017) Role of green tea extract (*Camellia sinensis*) in prevention of nicotine induced vascular changes in buccal mucosa of albino rats. International Medical Journal (1994) 24:120–125

Nag T, Ghosh A (2013) Cardiovascular disease risk factors in Asian Indian population: a systematic review. J Cardiovasc Dis Res 4(4):222–228

Nagpal M, Sood S (2013) Role of curcumin in systemic and oral health: an overview. J Nat Sci Biol Med 4(1):3–7

Naseem M, Khiyani MF, Nauman H, Zafar MS, Shah AH, Khalil HS (2017) Oil pulling and importance of traditional medicine in oral health maintenance. Int J Health Sci Res 11(4):65–70

Nathan C (2012) Fresh approaches to anti-infective therapies. Sci Transl Med 4(140):1–25

Navi F, Motamedi MHK, Fayaz F, Shabani M, Shams A (2014) Can 2% hydrogen peroxide-silver be an effective natural disinfectant in the dental office? J Oral Hyg Health 2:143

Nazir MA (2017) Prevalence of periodontal disease, its association with systemic diseases and prevention. Int J Health Sci Res 11(2):72–80

Negrato CA, Tarzia O (2010) Buccal alterations in diabetes mellitus. Diabetol Metab Syndr 2:3

Noble JM, Borrell LN, Papapanou PN, Elkind MS, Scarmeas N, Wright CB (2009) Periodontitis is associated with cognitive impairment among older adults: analysis of NHANES-III. J Neurol Neurosurg Psychiatry 80(11):1206–1211

Okamoto N, Morikawa M, Okamoto K, Habu N, Iwamoto J, Tomioka K, Saeki K, Yanagi M, Amano N, Kurumatani N (2010) Relationship of tooth loss to mild memory impairment and cognitive impairment: findings from the Fujiwara-Kyo study. Behav Brain Funct 6:77

Oklahoma Dental Association Patient's page (2014) The effects of oil pulling. J Okla Dent Assoc 2014:105–107

Oswal R, Charantimath S (2011) Herbal therapy in dentistry: a review. Innov J Med Health Sci 1:1–4

Pallasch TJ, Slots J (1996) Antibiotic prophylaxis and the medically compromised patient. Periodontol 2000 10:107–138

Palombo EA (2011) Traditional medicinal plant extracts and natural products with activity against oral bacteria: potential application in the prevention and treatment of oral diseases. Evid Based Complement Alternat Med 2011:1–15

Par M, Badovinac A, Plancak D (2014) Oral hygiene is an important factor for prevention of ventilator-associated pneumonia. Acta Clin Croat 53:72–78

Parolia A (2009) Oil hygiene. Br Dent J 207:408

Parveen A, Parveen B, Parveen R, Ahmad S (2015) Challenges and guidelines for clinical trial of herbal drugs. J Pharm Bioallied Sci 7(4):329–333

Passos VF, Melo M, Lima JPM, Marçal FF, Costa C, Rodrigues L, Santiago S (2018) Active compounds and derivatives of camellia sinensis responding to erosive attacks on dentin. Braz Oral Res 32:e40

Patrick DL, Lee RSY, Nucci M, Grembowski D, Jolles CZ, Milgrom P (2006) Reducing oral health disparities: a focus on social and cultural determinants. BMC Oral Health 6(Suppl 1):S4

Peedikayil F, Sreenivasan P, Narayanan A (2015) Effect of coconut oil in plaque related gingivitis - a preliminary report. Nigerian medical journal: journal of the Nigeria Medical Association 56:143–147

Petrovska BB (2012) Historical review of medicinal plants' usage. Pharmacogn Rev 6(11):1–5

Preet R (2013) Health professionals for global health: include dental personnel upfront!. Glob Health Action 6. https://doi.org/10.3402/gha.v6i0.21398.

Razak PA, Richard KM, Thankachan RP, Hafiz KA, Kumar KN, Sameer KM (2014, Nov-Dec) Geriatric oral health: a review article. J Int Oral Health 6(6):110–116

Reed D (2014) Healthy eating for healthy nurses: nutrition basics to promote health for nurses and patients. Online J Issues Nurs 19(3):7

Reynolds CM, Gray C, Li M, Segovia SA, Vickers MH (2015) Early life nutrition and energy balance disorders in offspring in later life. Nutrients 7(9):8090–8111

Rudolph MJ, Ogunbodede EO, Mistry M (2007) Management of the oral manifestations of HIV/AIDS by traditional healers and care givers. Curationis 30:56–56

Sanz M, Kornman K, Working group 3 of joint EFPAAPw (2013) Periodontitis and adverse pregnancy outcomes: consensus report of the joint EFP/AAP workshop on periodontitis and systemic diseases. J Clin Periodontol 40(Suppl 14):S164–S169

Sanz M, Beighton D, Curtis MA, Cury JA, Dige I, Dommisch H et al (2017) Role of microbial biofilms in the maintenance of oral health and in the development of dental caries and periodontal diseases. Consensus report of group 1 of the Joint EFP/ORCA workshop on the boundaries between caries and periodontal disease. J Clin Periodontol 44(18):S5–S11

Scardina GA, Messina P (2008) Nutrition and oral health. Recenti Prog Med 99(2):106–111

Segura-Egea JJ, Castellanos-Co Healthy L, Machuca G, López-López J, Martín-González J, Velasco-Ortega E (2012) Diabetes mellitus, periapical inflammation and endodontic treatment outcome. Med Oral Patol Oral Cir Bucal 17:356–361

Settineri S, Rizzo A, Liotta M, Mento C (2017) Clinical psychology of oral health: the link between teeth and emotions. SAGE Open 7(3):1–7

Sicca C, Bobbio E, Quartuccio N, Nicolò G, Cistaro A (2016) Prevention of dental caries: a review of effective treatments. J Clin Exp Dent 8(5):604–610

Singh A, Bharathi MP, Sequeira P, Acharya S, Bhat M (2011) Oral health status and practices of 5 and 12 year old Indian tribal children. J Clin Pediatr Dent 35(3):325–330

Singh P, Bey A, Gupta ND (2013) Dental health attitude in Indian society. J Int Soc Prev Community Dent 3(2):81–84

Sinha DJ, Sinha AA (2014) Natural medicaments in dentistry. Ayu 35(2):113–118

Sischo L, Broder HL (2011) Oral health-related quality of life: what, why, how, and future implications. J Dent Res 90(11):1264–1270

Sofowora A, Ogunbodede E, Onayade A (2013) The role and place of medicinal plants in the strategies for disease prevention. Afr J Tradit Complement Altern Med 10(5):210–229

Sooryavanshi S, Mardikar BR (1994) Prevention and treatment of diseases of mouth by gandoosha and kavala. Anc Sci Life 13:266–270

Spahr A, Klein E, Khuseyinova N, Boeckh C, Muche R, Kunze M, Rothenbacher D, Pezeshki G, Hoffmeister A, Koenig W (2006) Periodontal infections and coronary heart disease: role of periodontal bacteria and importance of total pathogen burden in the Coronary Event and Periodontal Disease (CORODONT) study. Arch Intern Med 166(5):554–559

Sravani K, Suchetha A, Mundinamane DB, Bhat D, Chandran N, Rajeshwari HR (2015) Plant products in dental and periodontal disease: an overview. Int J Med Dent Sci 4(2):913–921

Stein PS, Desrosiers M, Donegan SJ, Yepes JF, Kryscio RJ (2007) Tooth loss, dementia and neuropathology in the Nun study. J Am Dent Assoc 138:1314–1322

Stewart R, Hirani V (2007) Dental health and cognitive impairment in an English national survey population. J Am Geriatr Soc 55:1410–1414

Sudarshan GR, Vijayabala S (2012) Role of ginger in medicine and dentistry—an interesting review article. Southeast Asian J Case Rep Rev 1:66–72

Sultan Z, Zafar MS, Shahab S, Najeeb S, Naseem M (2016) Green tea (Camellia Sinensis): chemistry and oral health. Open Dent J 10:3–10

Swamy MK, Akhtar MS, Sinniah UR (2016) Antimicrobial properties of plant essential oils against human pathogens and their mode of action: an updated review. Evid Based Complement Alternat Med 2016:1–21

Tada A, Hanada N (2010) Opportunistic respiratory pathogens in the oral cavity of the elderly. FEMS Immunol Med Microbiol 60:1–17

Taheri JB, Azimi S, Rafeian N, Zanjani HA (2011) Herbs in dentistry. Int Dent J 61:287–296

Takeshita T, Kageyama S, Furuta M, Tsuboi H, Takeuchi K, Shibata Y et al (2016) Bacterial diversity in saliva and oral health-related conditions: the Hisayama study. Sci Rep 6:22164

Tardugno R, Pellati F, Iseppi R, Bondi M, Bruzzesi G, Benvenuti S (2017) Phytochemical composition and in vitro screening of the antimicrobial activity of essential oils on oral pathogenic bacteria view supplementary material phytochemical composition and in vitro screening of the antimicrobial activity of essential oils on oral pathogenic bacteria. Nat Prod Res 32:1–10

Taylor GW, Borgnakke WS, Graves DT (2010) Chapter 6 Association between periodontal diseases and diabetes mellitus. In: Periodontal disease and overall health: a clinician's guide. Professional Audience communications, Yardley, p 83

Thaweboon S, Nakaparksin J, Thaweboon B (2011) Effect of oil-pulling on oral microorganisms in biofilm models. Asia J Public Health 2:62–66

Türp JC, Spranger H (2016) Non-communicable disease and their significance for dental medicine. Swiss Dent J 126:473–489

Vaisocherová-Lísalová H, Víšová I, Ermini ML, Špringer T, Song XC, Mrázek J et al (2016) Low-fouling surface plasmon resonance biosensor for multi-step detection of foodborne bacterial pathogens in complex food samples. Biosens Bioelectron 80:84–90

WHO (2009) WHO monographs on selected medicinal plants, vol 4. http://www.who.int/medicine.docs/en/m/abstract/Js16713e

WHO (2010) WHO monographs on medicinal plants commonly used in the newly independent states. http://www.apps.who

Wiesner J (2017) Sensitization and allergies of herbal products. In: Toxicology of herbal products, Springer International Publishing, pp 237–269

Wolfram S (2007) Effects of green tea and EGCG on cardiovascular and metabolic health. J Am Coll Nutr 26:373S–388S

World Health Organization Monographs on Selected Medicinal Plants (1999) Volume 1. Available from: http://www.who.int/medicine docs/en/d/20.html

World Health Organisation (2017) Cardiovascular diseases (CVDs). http://www.who.int/news-room/factsheets/detail/cardiovascular-diseases-(cvds)

Wu T, Trevisan M, Genco RJ, Dorn JP, Falkner KL, Sempos CT (2000) Periodontal disease and risk of cerebrovascular disease: the first national health and nutrition examination survey and its follow-up study. Arch Intern Med 160:2749–2755

Wynn RL (2005) Aloe vera gel: update for dentistry. Gen Dent 53:6–9

Yadav K, Prakash S (2016) Dental caries: a review. Asian J Biomed Pharm Sci 06:01–07

Yoo S, Murata RM, Duarte S (2011) Antimicrobial traits of tea- and cranberry-derived polyphenols against *Streptococcus mutans*. Caries Res 45:327–335

Yuan H, Ma Q, Ye L, Piao G (2016) The traditional medicine and modern medicine from natural products. Molecules 21:559

Zaheer N, Shahbaz M, Athar Y, Arshad A, Zaheer U, Alam M (1994) Role of green tea extract (Camellia sinensis) in prevention of nicotine induced vascular changes in buccal mucosa of albino rats. Int Med J 24:120–125

Zi MYH, Longo PL, Bueno-Silva B, Mayer MPA (2014) Mechanisms involved in the association between periodontitis and complications in pregnancy. Front Public Health 2:290

Zougagh S, Belghiti A, Rochd T et al (2018) Medicinal and aromatic plants used in traditional treatment of the Oral pathology: the Ethnobotanical survey in the economic capital Casablanca, Morocco (North Africa). Nat Prod Bioprospect 2018:1–14

Printed by Printforce, the Netherlands